# Management Division 2024



Held at the 2024 AIChE Spring Meeting and 20th Global Congress on Process Safety

New Orleans, Louisiana, USA
24-28 March 2024

ISBN: 979-8-3313-1702-7

**Printed from e-media with permission by:**

Curran Associates, Inc.
57 Morehouse Lane
Red Hook, NY 12571

**Some format issues inherent in the e-media version may also appear in this print version.**



Printed with permission by Curran Associates, Inc. (2025)



AIChE
120 Wall Street, FL 23
New York, NY 10005-4020

Phone: (800) 242-4363
Fax: (203) 775-5177

www.aiche.org

**Additional copies of this publication are available from:**

Curran Associates, Inc.
57 Morehouse Lane
Red Hook, NY 12571 USA
Phone: 845-758-0400
Fax: 845-758-2633
Email: curran@proceedings.com
Web: www.proceedings.com

# TABLE OF CONTENTS

## 44c Transforming Diversity, Equity and Inclusion (DEI) Initiatives

*Lara Petrishchev*

Has remote work affected DEI initiatives? My goal is to dive into these types of questions as companies permanently evolved their work policies after the COVID-19 pandemic. Several employers are offering hybrid positions to remain competitive in attracting and retaining talent. Rather than a traditional DEI discussion, I am aiming to bring new perspectives based on modern issues. For instance, digital transformations for industry workers have affected how often they may interact with others on-site.

Since the issues around DEI have transformed over time, DEI metrics must continuously be evaluated for all types of workers to ensure a healthy work culture. This session will cover best practices and insights to attract, retain, and develop talent for a sustainable workforce. Audience participation is encouraged.

## 44d Attract and Retain Talent, A Timeless Process

*Markus Scheller*

Hiring people seems to be an easy process, but are the new organizations and the fresh employees really fit for a successful collaboration? Does the company culture support and inspire the envisioned partnership? The responsible manager and his/her team creates the required, needed and expected atmosphere for highly effective teamwork not only as a part of the hiring process, but also for the longer lasting successful job. But does every employee have a clear view of his / her needs and what are the elements of a healthy life balance? Therefore, the interesting questions are: what creates a motivating, nourishing environment for employees and employers?

## Strategy to Craft Unique and Customized Careers within the Organizational Framework

*Tianxing Cai*

In an age of rapid technological advancements and shifting professional landscapes, the traditional career path is evolving. This presentation explores the strategy to empower individuals in crafting unique, customized careers within the organizational framework. The presentation is about unleashing one's inner innovator and applying entrepreneurial thinking to transform one's career. We delve into the key components, including creativity, risk-taking, and intrapreneurial leadership. The presentation also offers practical strategies and tools for participants to embark on their own intrepreneurial journey. From identifying opportunities for innovation to building cross-functional teams and garnering support from leadership, attendees will leave with actionable steps to innovate and design their customized career paths. Moreover, we explore the benefits of embracing an intrepreneurial mindset, including increased job satisfaction, career fulfillment, and the potential to drive organizational change from within. This presentation inspires and equips individuals to reimagine their professional journey. By embracing intrepreneurship, attendees will learn how to innovate, take calculated risks, and design a career that aligns with their passions and aspirations, ultimately redefining success on their own terms in an ever-changing world of work.

## 24b Integration of two Process Safety Management Systems

*Wing Yee*

When two global companies with their own Process Safety Management (PSM) merged, the journey of integrating and harmonizing of the system began. This presentation will demonstrate the process we went through and challenges/opportunities we have faced as we migrate personnel, documentation, management system together. At the same time, we utilized this opportunity to make our PSM better and create a safer environment

## 24c Integrating Process Safety into Organizational Leadership

*Jeremy Lancaster*

Article Many organizations have plateaued in achieving the desired process safety results and are faced with an exit of strategic leaders (all levels). Frontline leaders may have less experience at leading the organization and require an acceleration of learning, followed by a demonstration of understanding to ensure success in process safety. Leaders at times *"think and say they are doing great, until a process safety event occurs"*. Process safety competencies is one approach which goes beyond basic Process Safety Management (PSM) compliance to help ensure culture is consistent and process safety risks are reduced. The concept is built on a competency matrix which identifies the leader positions, skills, and knowledge, with proficiency demonstration. This model can be adopted and modified to fit any organizational structure who desires to make a step-change improvement. A barrier analysis to identify what might prevent the knowledge transfer and sustaining the program will be required to supplement your competency program launch.

Process safety competency is intended to avoid loss of containment of chemicals with potentially serious consequences.

Human factors, which have been studied for decades, also play a key role in the process of safety competency success within any organization. Understanding and using basic human factors concepts helps leaders understand why people do the things they do and help workers perform the task correctly. Simply put, Human factors consider the human elements of any task and are critical to the success of process safety performance. Every organization spends time on human elements following an incident during the investigation process. Process safety competencies integrate the desired outcomes and requirements at the forefront as process safety leaders demonstrate their responsibilities.

Actions and tools available to address human factor scenarios include critical task analysis, job-aids (checklists), managing competency in the organization, and eliminating the error-likely scenarios with the highest consequences.

Competency assessment can be accomplished through a variety of techniques including written exams, observations and testing, job shadowing, verbal questioning, and simulations.

A monitoring and management system to keep up with the ever-changing cycle of leader positions will be necessary to ensure process safety competency continues. Process safety event outcomes is one measure, but increased field presence, increased awareness with key leaders, and increased engagement in process safety work processes are all methods which can be monitored to help ensure the long-term process safety success of an organization.

## 131a Presentation of 2024 Young Leader Award to Dr. Polina Ware

*Cristina Thomas*

Cristina Thomas, Chair of the Management Division's Awards Committee, will present the 2024 Young Leader Award to Polina Ware of the Rogers Corporation. The award is sponsored by the Chemours Company. This award recognizes an outstanding leader under 40 years old who has made a significant contribution to the success of their organization by advancing a scope, project, unit, plant, division, technology, strategy, or company that made an impact to the practice or field of chemical engineering.

## 131b Strategizing AI Integration: Advanced Frameworks for Implementing and Adaptation of Generative AI and Digitalization in Specialty Chemical Industry Sector Digitalization

*Polina Ware*

This presentation delves into the strategies for embedding Generative AI into the core of technology organizations, underscored by scientific principles and practical methodologies. First the latest advancements in Generative AI, focusing on its algorithmic foundations, potential for innovation, and implications for data management and analytics are discussed. Then, the presentation transitions into examination of the digital transformation process, use cases and driving cultural change and adaptation including strategies for change management practices that are essential for successful integration of this technology. This talk addresses overcoming challenges such as data compatibility, scalability and most importantly gaining user acceptance and buy-in across all levels of organization. This part of the presentation will address the human and organizational aspects of AI integration: managing change, focusing on staff training, leadership in technological adaptation, and fostering a culture of continuous innovation and ethical AI use. This presentation aims to foster a dialogue among R&D leaders about the future of technology development, centered around the intelligent and sustainable integration of Generative AI.

## 131c AI Driven Advancements in Industrial Biotechnology: Transformative Solutions for Sustainable Innovation

*Sajith Wimalaratne*

AI-driven advancements in industrial biotechnology are revolutionizing the landscape of bio-based manufacturing processes. By harnessing the power of generative AI, researchers and engineers can optimize fermentation, protein engineering, metabolic engineering, and bioprocess design, leading to increased efficiency, higher yields, and accelerate the products to market.

In this presentation, we will address.

1. How AI and deep tech is already being integrated into the biotech sector to accelerate commercialization.
2. The challenges that the sector is facing and the potential solutions.
3. How much of an issue is the safety and how to build trust with the users and what can the leaders do.

## Picking the Right AI Bets

*Zee Hussain*

A top reason why 70% digital transformation projects and over 90% startups fail is the inability to answer the question - Is this a real problem, and why should we solve it over others? This presentation covers three things, in the context of AI projects/products.

* Finding the right problems to solve using a human centric approach

* Validating solutions before actually building them

* Finally, picking the right ideas to pitch/back, using an analytical framework.

Though the focus will be on AI, the concepts covered are universal, translating across technologies and domains.

## 5a Leadership in Action Via Aiche Volunteerism

*Christina Thomas*

Interest in Sustainability has grown exponentially and with it the demand to act has materialized. Thanks to the growth in volunteerism in Sustainability, we now have many government agencies, industries and non-profit organizations working together to address the challenges that come to realize a sustainable world. Professional societies related to the chemical industry have taken numerous and important steps to inform its memberships including on matters of public affairs. During this talk, I will illustrate how the AIChE PAIC (Public Affairs and Information Committee) is working to take important steps towards Sustainability.

## 5b Profitable Decarbonization: How Digital Twins Increase Efficiency and Reduce Emissions

*Sterling Allen*

Industry has competing interests: continue to deliver the products people rely on while simultaneously decarbonizing to fight climate change. What most are not aware of is that initial decarbonization steps can be taken with tools that provide significant $CO_2$ reduction while also increasing existing process efficiency. Those efficiency gains will not only pay for the digital twin tools at the heart of the solution, but will also positively impact bottom line performance.

In addition to discussing the underlying digital twin technology, the speakers will address their recent partnership and its ability to address customer needs.

## 5c Sustainability 2.0

*Leah Beckworth*

The concept of sustainability at UCOR far exceeds the bounds of how it is traditionally defined. While we certainly work hard to reduce our carbon footprint and greenhouse gas emissions as a part of our Clean Up to Clean Energy Mission, we have grown our sustainability paradigm to evolve into a much more holistic perspective. Through a sustainability lens we asked ourselves –

- Are our business systems sustainable?
- Is our workforce sustainable?
- Is our community sustainable?

This session will specifically explore where worker feedback and challenging our sustainability status quo led to UCOR establishing a robust behavioral and mental health program called Navigate.

Navigate seeks to educate, engage, provide resources, normalize mental health conversations, and invest in people and their wellbeing. The session will explore UCOR's behavioral and mental health strategic plan which includes the vast array of "tools in the toolbox" required to impact the program's targe population.

Through Navigate UCOR has enabled integrated, holistic approaches that ensure viable, long-term behavioral and mental health solutions are achieved.

# AUTHOR INDEX